A cooper's berth.

T0174475

Coopers
and Coopering

Ken Kilby

Published by Shire Publications Ltd,
PO Box 883, Oxford OX1 9PL, United Kingdom.
PO Box 3985, New York, NY 10185-3983, USA.
E-mail: shire@shirebooks.co.uk   www.shirebooks.co.uk

Copyright © 2004 by Ken Kilby.
First published 2004.
Transferred to digital print on demand 2014.
Shire Library 426. ISBN-13: 978 0 74780 584 7.
Ken Kilby is hereby identified as the author of this work in
accordance with Section 77 of the Copyright, Designs and
Patents Act 1988.

British Library Cataloguing in Publication Data:
Kilby, Ken
Coopers and Coopering. – (Shire Library; 426)
1. Coopers and Cooperage – History
I. Title
674.8'2
ISBN 0 7478 0584 9.

Cover: The author's father with his day's work, c. 1913. Bung staves received a lot
of punishment, hence these casks have all had bung staves repaired.

ACKNOWLEDGEMENTS
I would like to acknowledge the kind and generous help given me by Mr Paul Jupe; Mr
James Kilby; Mr Peter John of AP John, Coopers, Australia; Mr Harry Hobbs of Hamilton,
Ontario, Canada; Mr Craig Dunn of Baltimore, USA; Mr Alexander Robertson of Great
Yarmouth; and many others with whom I have come into contact during my research.
  Many thanks to Wadworth & Company Limited for supplying the photograph on the
front cover, and to the following for their kind permission to use the illustrations indicated:
A.P. John & Sons, Coopers, pages 11 (bottom) and 23; Bembridge Maritime Museum, page
39 (top left); Champagne Mercier, pages 52-3; Braumeister Craig Dunn, pages 57, 58;
Cadbury Lamb, page 58; the Guinness Archive Library, page 7; the Langdale Estate,
Cumbria, page 40 (top); the Mary Rose Trust, page 38 (right); the Niagara Falls Public
Library, page 3; the Warshi *Vasa* and Anders Franzen, page 37 (bottom); Watford Museum,
page 29; the Weald & Downland Open Air Museum and Diana Zeuner, page 26.

Printed and bound in Great Britain.

Contents

In 1886 Carlisle Graham, an English cooper, was the first person to ride the Niagara Falls and Rapids in a 'barrel'. He repeated this feat on a number of other occasions. In 1901 Anna Edson Taylor successfully challenged the Falls in a conventionally shaped wine cask, demonstrating its immense strength (and her incredible nerve).

The barrel: an invention of distinction

If you were asked to name the greatest invention of all time you would probably say the wheel, or the steam engine, or the internal combustion engine, or perhaps one of a number of other inventions that might warrant this distinction. Few if any would consider the humble barrel. Yet, for thousands of years most commodities were moved, shipped or kept in barrels. Why? Because barrels were exceptionally strong, with hoops binding the joints into the form of a double arch; because they were in themselves wheels, a means of movement at a time when power was dependent upon the muscles of man or beast; and because certain goods actually benefited from being in a barrel. Without the barrel most goods would have remained right where they were made, or not have been made at all. Few inventions have stimulated such enormous, widespread demand over so many centuries.

On the left is a small 4¹/₂ gallon pin (20.5 litres) and behind that an 18 gallon kilderkin (82 litres). In the centre is an encrusted amphora which would probably hold 6 gallons (27 litres). On the right is a 54 gallon hogshead (245 litres) and at the back a 108 gallon pipe (491 litres).

The original barrel-shaped drum from Ancient Egypt is in the Metropolitan Museum of Art, New York, and somewhat dilapidated. The drum in this picture was made like that first barrel-shaped vessel.

Wooden casks were made in thousands of small coopers' shops dotted all over Britain, where coopers and the sons and grandsons of coopers toiled away at their blocks. There are few jobs as physically demanding as coopering, amidst the noise of hammering and the smoke of firing. Few jobs require as much skill as the making of a cask. Coopers served a seven-year apprenticeship, swinging axes and adzes, drawing *long-knives* and slamming hammers, eventually with such precision that they could guarantee to make a watertight vessel in any one of a wide variety of shapes and sizes. Although many people call any sort of cask a barrel, in fact a *barrel* is a cask that holds 36 gallons (164 litres). Smaller casks are *kilderkins* of 18 gallons (82 litres), *firkins* of 9 gallons (41 litres) and *pins* of 4¹/₂ gallons (20.5 litres). Larger casks are *hogsheads* of 54 gallons (245 litres), *puncheons* of 72 gallons (327 litres), and *butts* and *pipes* of 108 gallons (491 litres). A *tun* is of 225 gallons (1023 litres) but is now seldom used.

Tomb paintings indicate that straight-sided wooden buckets bound with wooden hoops were made in Egypt as early as 2690 BC. Vessels described as barrels are mentioned in the Bible and by the Greek historian Herodotus writing in the fifth century BC. The first recorded barrel-shaped vessel was a drum, made in Ancient Egypt between 1580 and 525 BC. Yet casks were not used for trading at that time, most probably because of the lack of suitable timber in Egypt. For thousands of years the 6 gallon (27 litre) clay amphora was the container used for trading purposes. It could hold wine, oil or fish but had to be carried aboard ship and stacked upright and tight against its neighbour for fear of breakages.

An excavated Iron Age silver-fir cask at the Mittelrheinisches Landesmuseum in Mainz, Germany.

Part of a Roman silver-fir cask from excavations at Canary Wharf, London, showing the neatness of the chiming.

According to the Roman historian Pliny the Younger (61–113 AD) the wooden cask was invented in Cisalpine Gaul, the region of the province of Gallia south of the Alps. Broken potsherds of amphora become scarce in widening distances from this area, suggesting their replacement by wooden casks. These first casks were made of quartered silver fir and have been unearthed in England, where, after being emptied of wine, they were used to line wells. The wooden cask would stimulate trade and shipbuilding as never before: a veritable revolution.

The author's father with his day's work, in about 1913. Bung staves received a lot of punishment, hence these casks have all had bung staves repaired.

Have you ever wondered why Guinness has such a distinctive flavour? This photograph shows coopers at the Guinness brewery in Dublin charring the insides of their casks. Casks for maturing whisky were also charred in this way.

Making a barrel

The cooper keeps his tools on his bench or propped up beside it. This is what his bench is for; he works at the block. The tools are kept razor sharp so that they will cut through a piece of *flag* (rush) with only the force of their own weight, without leaving a rough edge. They wear so that they sympathise with the wood. Raw linseed oil is rubbed on to the working surfaces to reduce friction.

The timber, cut to the appropriate length, width and thickness, is known as a *stave*. First it is carefully inspected for blemishes and to determine which way it will bend more easily without breaking. It is then *dressed*. To do this, the cooper first holds the stave across his block and puts a rough shape to it with his axe; this is called *listing*. The stave is then backed and hollowed out with long-knives, the longer staves on the block, the shorter ones on a *horse*. Lastly it is jointed on a *jointer*, a long

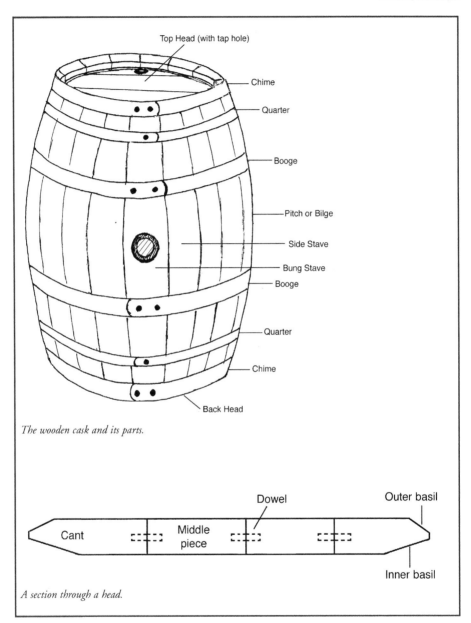

The wooden cask and its parts.

A section through a head.

upturned plane, in order to put the *shot*, or angle, on the shape of the stave, corresponding to the radius of the cask throughout its length. The cooper also looks carefully down each jointed edge to see how much *height* (belly) he is putting into the cask, which he judges with his practised eye. As these staves fit together in the cask, the jointing must be so accurate that the butt joints will not allow leakage under as much as 30 pounds pressure per square inch.

When sufficient staves have been dressed, they are *raised up* in a *raising-up hoop*, which is filled to give the cask the correct capacity; the cooper will judge this to within a pint. It is never wise to put a soft stave next to a tough one and a good cooper sorts out his staves very carefully. With the staves

Backing the stave on the block.

Hollowing out on a horse.

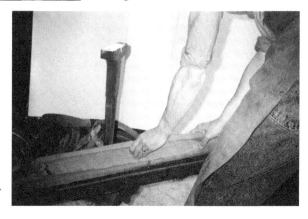

Jointing – the stave is pushed down the jointer plane.

Top left: The stave after listing.
Top right: The stave after backing.
Above: The stave after being hollowed out.
Right: The dressed stave after jointing. The dotted line indicates the amount of height (belly).
Left: The cask raised up ready for firing.
Below: A horse (shingle horse; known in Ireland as a mare).
Below left: The jointer.
Below right: The block with a block hook.

The preparation of the staves and the raising up of the cask.

Listing the stave.

Below: *Raising up a large cask. Note the flower or figuration on the staves in the foreground. This is where the medullary rays come to the surface, indicating that it is perfectly quartered oak. This is often seen on the best furniture. A cooper would make his own metal clips with old pieces of hoop iron.*

firm in the raising-up hoop, the cooper puts a *truss hoop* called the *first runner* (or, up to the eighteenth century, the *gathering hoop*) over the cask and tightens it up. This again gives an indication of how much height the finished cask will have. A metal *booge hoop* is then driven down on to the cask and the truss hoop removed.

Stout and *extra stout* casks (with staves of 1¼ and 1½ inch thickness – 3.2 or 3.8 cm) need to be immersed under a steam bell or steeped in boiling water for half an hour to soften the timber and to make it more pliable. *Slight* casks need only to be put over a *cresset* of burning shavings until they are warm through to the outside and they are ready to bend. Then, with a shout of 'Truss oh!', two coopers or a man and his apprentice will team up. The first runner is thrown back over the cask, still heated over the cresset, and the hammering with trussing adzes will begin. Smaller truss

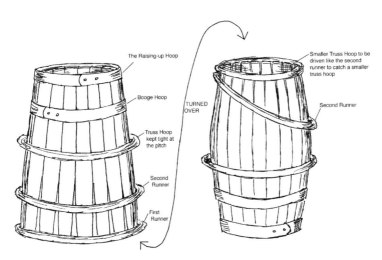

The Raising-up Hoop

Booge Hoop

TURNED
OVER

Truss Hoop
kept tight at
the pitch

Second
Runner

First
Runner

Smaller Truss Hoop to be
driven like the second
runner to catch a smaller
truss hoop

Second Runner

Above: *The firing process. The fire is kept burning in the cresset inside the cask.*

Right: *The firing (bending of the staves) before driving on the dingee.*

hoops will be driven down until the open end of the cask is pinched in. The cask is then turned over.

Speed is of the essence. The longer it takes, the harder it will be; and you can imagine the exasperation of the cooper, his eyes smarting from the smoke and running sweat, when a hoop is stubborn or, worse, when one breaks under the strain. The truss hoops are driven down on one side until a smaller one can be caught on over the staves. This process is repeated until the staves are completely bent and the cooper can catch a *dingee hoop*, the same size as a raising-up hoop, on the other end. The hoops on the *pitch* (belly) must be kept tight to stop any staves breaking. A cracked stave is called a *duck* – a costly occurrence dreaded by coopers, especially when they are on piecework. The fire is kept burning in the cresset so that the staves acquire *set*, and then, if the hoops are removed, the staves will retain their bend. At this stage, having been fired, the cask is called a *gun*.

The cask now has to be *chimed* and a *chiming hoop*, a hoop slightly bigger than the raising-up hoop which it replaces, is driven into position. The cask is leaned against the block and a *bevel* or slope is cut on the ends of the staves with an adze. To make sure the *chimes* (the ends of the

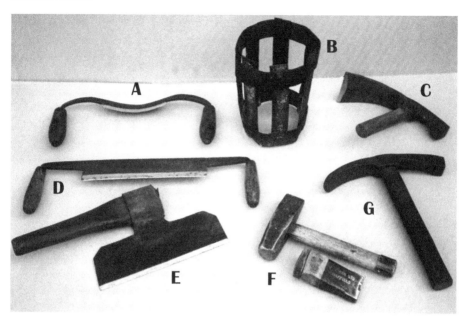

A, hollow knife. B, cresset. C, adze. D, backing knife. E, axe. F, hammer and driver. G, trussing adze.

staves) are square, the cooper runs a *topping plane* round the top of the chime. If necessary, he then goes round again with his sharp adze.

To make the inside of the chime perfectly curved so that a groove can be cut into it to take the head, the cooper uses a *chiv*, resting the cask between his knee and the block. On casks of unusual size this can be done with a *jigger*, a one-handled *drawknife*, but it requires considerable

A, knocker-up (devil's tail). B, white cooper's inside shave. C, pail shave. D, thief auger. E, bung auger. F, flagging iron. G, diagonals. H, rat-tail file. I, file. J, rasp.

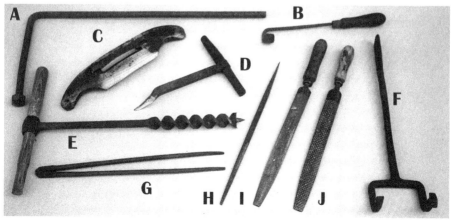

Using an adze to cut the bevel on the top of the staves – chiming the cask.

skill. A *rining adze* could be used on very large casks. The *croze* is the tool that cuts the groove. It is swung round the inside of the chime in the same way as the chiv, care being taken to keep the depth constant. Cutting the groove on a small cask or bucket is done on a horse, keeping the vessel upright. With one end of the cask chimed, it can serve as a *case*, the staves being used to repair other casks; some small brewery cooperages used to buy cases rather than fire their own casks.

The other end of the cask then has to be chimed. Before the cooper cuts the groove he checks the potential capacity with his *diagonals*, two lengths of metal rod hinged together, so that he can make any necessary adjustments to the groove or the head to rectify an error. This is seldom necessary. It only remains then to smooth the inside of the cask with an *inside shave* so that it can be sterilised effectively and no bacteria will lurk in any rough area to turn the beer sour. As coopers often bark their arms shaving out small casks, some use extended inside shaves to prevent this, and also to save their backs.

Shaving the inside of the chime with a chiv.

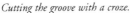

Cutting the groove with a croze.

A, bick iron. B, heading board. C, heading knife. D, swift. E, heading vice. F, dowelling stock. G, compasses.

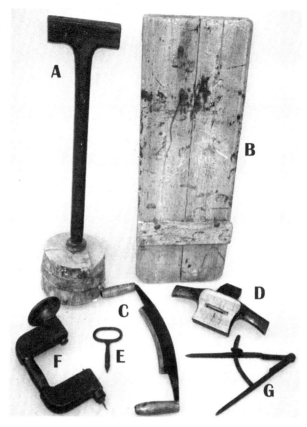

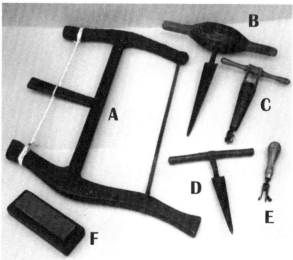

A, bow saw. B, tapered auger (bung-hole). C, American tapered auger. D, tapered auger (tap hole). E, race. F, oilstone (carborundum).

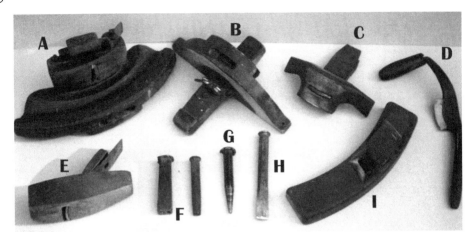

A, chiv. B, croze. C, inside shave. D, jigger. E, stoup plane. F, rivet moulds. G, punch. H, cold chisel. I, topping (or sun) plane.

At this point the cooper usually makes the permanent hoops for the cask. He buys his hoop iron already cut to the appropriate sizes and splayed to fit the curve of a cask. The cooper holds a length of hoop iron round the cask just above where it is to be driven and places his thumb on the spot where he will rivet and join the two flaps. Taking it to the *bick iron*, he hammers a rivet through each flap and burrs it over.

A, dry cooper's adze. B, downright. C, extended stoup plane. D, round shave (smuggling iron). E, buzz. F, herring cooper's flincher. G, scraper.

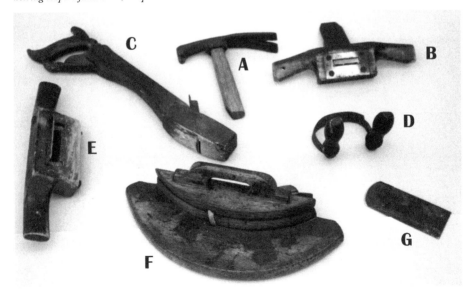

Boring dowel holes on the cant of the head using a dowelling stock.

Sometimes he will have to bruise more or less splay on to a hoop with the *nose* of his hammer.

Next the cooper makes the heads. To find the radius of the required head, he works his compasses round the groove until they fit exactly six times. He then selects his timber and joints the pieces of heading on his jointer. Holding them together up to the light, the cooper checks the joints and places them together on his *heading board*. He takes his *dowelling stock* and, holding it against his stomach, bores the dowel holes. He makes his own dowels out of a piece of stringy American red oak and fits flag between the joints before inserting the dowels and tapping the pieces of heading together. Bending over the heading board, he shaves both sides of the head with his *swift* and, after describing a circle with his compasses, he saws round the head with his *bow saw*.

Dowelling together the head. Note the flag in the joint.

Sawing round the head with a bow saw.

Below: *Shaving the head smooth with a swift on a heading board.*

Cutting in the head with a *heading knife* requires considerable skill. The cooper holds the head firm between his body and a notch in the block. He cuts the outer *basil* first. Next he marks again with his compasses. Since the inner basil is much bigger, he uses his axe to chip off some of the surplus wood. He then goes round the head with his heading knife, cutting to the compass mark but leaving a little extra on

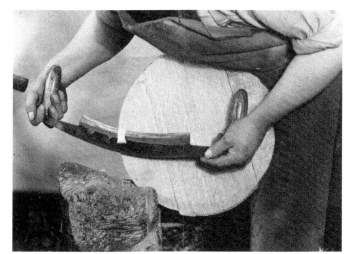

Cutting the outer basil of the head with a heading knife.

Shaving out the inside of the cask.

the cants to allow for squeezing of the joints. When he has made both heads he makes ready to fit them into the cask. To do this, he slackens the hoops on one end and forces the head he has chosen to be the back head into the cask. He inserts a length of flag around the groove and then puts one side of the head into the groove before turning the cask over so that the rest of the head will drop into position. He taps the head home with the handle of his hammer and he then turns the cask over and tightens the hoops. To pull the top head up into position, he uses a *heading vice*, which is screwed into the head in the position of the *tap hole*. Alternatively he might bore a *bung-hole*, through which he inserts a *knocker-up*. It is possible to get the head up into position by opening up

Above left: *Pulling the top head into position with a heading vice.*
Above right: *Buzzing down.*

a stave joint and using a flat piece of hoop iron to lever up the head.

The joints on the outside of the cask are now levelled with a downright. In order to give a smooth finish to the cask, the cooper uses a *buzz*, which is really a turned scraper held for leverage in a wooden holder between two handles. Finally the cooper hammers the hoops home with a hammer and *driver*. On larger casks the cooper will use an eight-pound sledgehammer, employing a *maul* on the chime hoop. When this is finished the head will ring like a bell.

Boring the bung-holes and tap holes is usually done by the odd-job man. To find the centre of the bung stave, he measures from each end of the cask with a piece of flag and chalk. After boring the hole with an auger he takes a tapered auger round the hole before screwing in the brass bush, or alternatively expanding a bush into the stave. Before brass bushes came into use in the twentieth century bung-holes and tap holes were burnt with red-hot tapered irons to seal the fibres and prevent rotting.

Timber

For *dry coopering* the accent was on cheapness and any soft wood or old staves would serve for fruit, potatoes, seeds or ironmongery. In *wet* and *white coopering*, on the other hand, the kind of wood used and the way it had been converted was of paramount importance.

The white cooper used mostly Memel oak, English oak, beech and chestnut for water, milk, butter or cheese vessels, where care had to be taken in case the wood imparted a taste to the contents. French wine coopers liked to use oak grown in the same soil as the grapes. Their casks were always charred and blistered inside to aid the maturing process.

Until the Second World War Memel oak was used almost exclusively for beer casks. It grew along the many rivers flowing down into the Baltic in forests that were predominantly fir, a faster-growing tree which caused oaks to be drawn up straight in a fight for light and air, until the 'nurse' trees were cut down. In this way Memel oaks developed with a near-perfect grain, free from knots, unlike English oaks, which grow singly and awry. Russian peasants, unable to follow their usual occupations in the winter, would fell the trees while the sap was down, remove the crown (branches) and bark and cut the trunk into stave lengths. The longest staves were pipe size, 5 feet 6 inches (168 cm). The lengths were then split in half with two-handed axes, or into wedges if the log was stouter than 2 feet 6 inches (76 cm) in diameter, and then split into billets radially. The woodsman marked the timber with a bow string rendered with charred timber, which he twanged on the oak. He then worked to the line with his two-handed axe, reducing the length of wood to a parallelogram in section. His skill was such that very little needed to be done when the staves were finished off on a large plane operated by two

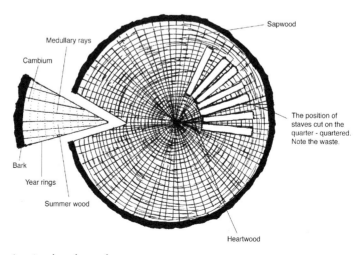

A section through an oak tree.

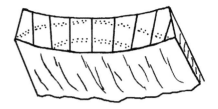

On the left is a section through a quartered stave. On the right is a section through an improperly cut stave that would be porous and warp.

men, one pulling, the other pushing. Staves were shielded from the wind with branches before being transported to the railhead and thence to the port. Workmen called 'sworn brackers' sorted the staves into best-quality *crown*, and second-quality *brack*, ploughing a mark in each with a scribing tool. Full-size staves were 3 inches (7.6 cm) thick and 6 inches (15 cm) wide. They were sold by the mille of 1200 pieces.

Since Memel oak was split with the grain it was called cleft oak. The timber was always perfectly quartered, so that the medullary rays, which radiate from the centre of the tree and form impervious layers that tend to keep their shape during seasoning, run across the staves as in the diagram above. The year rings, on the other hand, are porous. English oak can seldom be cleft like Memel oak as the grain is not sufficiently straight. It is often sawn so as to give short-grained staves but not well-quartered ones.

Persian oak was used after the Second World War, when Memel oak became unobtainable. However, this oak suffered from internal cracking of honeycomb form, caused through rapid drying in a hot climate, and twenty per cent waste was normal.

American red oak has an excess of acid-tasting tannin and casks made of it were often lined with a rubber or plastic solution. This oak was used by the Guinness brewery for their casks, which were afterwards *pompeyed*, that is charred inside. American white oak is used extensively for wines and spirits.

Timber needs to be seasoned, which is the process of drying it right through. Traditionally this was achieved by letting it weather for up to five years in stacks through which the air could circulate freely. An ordinance put before the Lord Mayor of London in 1488 explains what would happen if unseasoned timber were to be used:

> That where grete deceit and untrowthe daily…by the means of makyng barells…of sappy andd grene atymbaer for lack of serche and correction thereupon to be hadd and done…which…of necessite must shrynke and…lacke of their true and just measure that they ought to conteyne.

Preparing a cask for beer or wine would be a problem on which the advice of the cooper would be sought. In order to neutralise the acid-tasting tannin in the oak, the cask has to be soaked in a solution of salt and sodium carbonate before being filled with beer or wine. This is called 'pickling the cask'. Casks must have a sweet *nose*, and they are *snifted*

Timber in the process of being seasoned at the cooperage of A. P. John & Sons, Taminda, Australia.

before being *racked* (filled). If found to be only very slightly sour, they are left to soak in a sodium solution. After use, casks must be cleaned out with steam and hot *liquor* (water) and, if they are not to be used for some time, a quarter of a pint (142 ml) of sodium bisulphite solution is poured into them. They are then corked and pegged and shaken, while the outside is periodically swilled to keep the casks from drying out and the joints from opening up. They can be left empty for many months until required and will smell very sweet, needing only to be soaked and washed.

Ancillary tradesmen

Hoopers

The making of hoops for casks had been a specialist woodland trade for many centuries and was carried on in Sussex, the Midlands and Furness. Hoops were exported to Jamaica for sugar barrels, a trade which ended at the beginning of the twentieth century. The hoops were mostly of hazel, cut and split with an axe and trimmed with a drawknife set at right angles to the handles and called a *spokeshave*. The wood was then steeped in water to make it pliable and coiled on a horse, a frame of upright pegs. Coopers would make them to size by notching and overlapping the ends, binding and nailing them. Standard sizes of hoops were made by nailing them up within a stout ash hoop, which acted as a mould. One man could make between four hundred and five hundred hoops a day.

Rush gatherers

Rivermen like Metcalf Arnold, whose yard was on the Ouse at St Ives in Cambridgeshire, next to the Ferryboat Inn, used to harvest rushes, wading into the river and cutting them deep near the base. He then spread them out to dry in the sun. The best he called *coopers* and the rest he sold for mat-making and thatching.

Rush, or flag as it is known to coopers, swells quickly when it comes into contact with water. It was inserted into joints so that, if the empty cask were to dry out and the wood shrink, causing the cask to leak when refilled, the rush would swell and *take up* until the wood itself swelled to seal the joint.

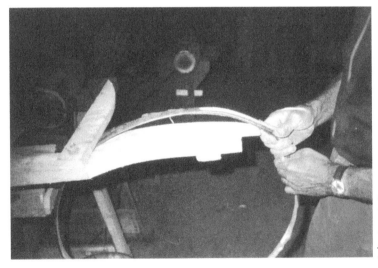

The cooper uses a hoop clamp to hold the hoop firm when joining the ends of the hoop for the necessary size.

The village cooper

From the earliest times the demand for coopered vessels was widespread. Small, often one-man cooperages sprang up in market towns and villages. Very little capital was required – a garden shed, a set of tools and old casks for the raw material.

In medieval times newly married couples would have to pay a visit to the cooper, someone with a name like Walter le Cuver, soon to be anglicised into the surname Cooper. They needed buckets for water and for milking, a corn bucket for feeding the chickens and maybe a stable bucket. They needed bowls for washing and a dolly tub for laundering. They needed pickling vats and storage vats for flour, oats and salt, and a churn for butter-making, tubs for cheese, and casks and tubs for home-brewed ale. Many of these casks would last a lifetime, valued possessions that were recorded in wills.

Village coopers had to be quite versatile and some became very skilled. The better ones had served apprenticeships in the city but, perhaps having fallen foul of the guild, chose to continue their trade in the country. They might have a stall at the local market town or pay a toll to sell in the city and, in time, their wares would include polished fruit bowls, wood bins and, later, coal scuttles. A very popular item in later years was the jardinière, the tub that stood on a table by the window in the front room and held the aspidistra.

The decline of the village cooper came about during the nineteenth century with the manufacturing of cheap galvanised and enamelled buckets, bowls and jugs. In addition the village cooper had often undertaken work for small breweries, keeping their casks in repair. This work dried up as small breweries were swallowed up by larger ones employing their own coopers, or they found it wiser to employ coopers from a larger cooperage of repute to undertake their repair work.

Coopers at work. Sketches from Pyne's 'Microcosm', 1802.

Above: *Various buckets. (Back row, left to right) Ice cream; plate; oval water. (Centre, clockwise) General purpose from eastern Europe; ship's; ice bucket; nineteenth-century water; bigger bottom for rough ground; stable. (Front row, left to right) Soft wood general purpose; milking pail; corn – for feeding chickens.*

Well buckets were the province of the village cooper. This one was made by the author for the treadwheel-operated well at the Weald & Downland Open Air Museum in West Sussex.

Commercial directories record the gradual demise. In Bedfordshire fifteen coopers were listed in 1853, fourteen in 1861, nine in 1871, three in 1894 and one in 1910. In Hertfordshire there were twenty-eight in 1864, twenty-three in 1878, six in 1899 and, in 1902, five.

William Dalley of Turvey, Bedfordshire, a ploughwright, cooper and carpenter, drawn from life, April 1830.

In the decades during which their numbers were dwindling coopers began to take on other work to survive. A famous sign over the cooper's shop at Hailsham in East Sussex in the early twentieth century ran:

As other people have a sign,
I say, just stop and look at mine!
Here Wratten, cooper, lives and makes
Ox bows, trug baskets, and hay rakes.
Sells shovels both for flour and corn,
And shauls, and makes a good box churn,
Ladles, dishes, spoons and skimmers,
Trenchers too, for use at dinners.
I make and mend both tub and cask,
And make 'em strong, to make them last.
Here's butter prints and butter scales,
And butter boards, and milking pails.

The last cooper in Bedfordshire.

N'on this my friends may safely rest –
In serving them I'll do my best;
Then all that buy, I'll use them well
Because I make my goods to sell.

The last village cooper in Bedfordshire, at Carlton, closed his doors for the last time in the 1970s, having survived for so long by making flower tubs from old whisky casks.

Coopering in the cities

The expansion of trade, particularly under the Tudors, stimulated demand for casks. In the cities coopers came together in guilds, primarily in order to create a 'closed shop', as in Coventry, where a cooper 'shall not occupie any shop within this Citie oneless he agre with the cowpers of this citie'. Upon completing his apprenticeship the cooper became an enfranchised freeman of the city. Developments during the sixteenth century also had a profound effect on coopering. Breweries were established and grew fast. By 1591 in London there were 'twenty great brew houses between Fleet Street and St Catherine's' and they tended to attract many of the best coopers. These were wet coopers, and they were offered higher remuneration and some relief from the monopoly of the guilds.

The Coopers' Company fought to maintain the independence of this branch of coopering and insisted that brewers bought their casks from independent cooperages. They took their case to Parliament, where they bribed the Lord Chancellor with half a butt of malmsey in 1533, the Lord Chief Justice with twenty gallons of sherry in 1561 and the Speaker of the House with a runlet of ten gallons of sack in 1562. However, although coopers managed to get Parliament to pass laws in their favour, the brewers flagrantly disregarded them. Perhaps they paid bigger bribes. So vigorously did these breweries proliferate that, in 1751, when Thrales brewery was being sold to Barclay, Dr Johnson commented, 'Sir, we are

The coopers of Benskin's brewery, Watford, in the 1950s. The cooper standing behind the left-hand seated man is the author's uncle, William, wearing his old, battered trilby.

not here to sell a parcel of boilers and vats, but the potentiality of growing rich beyond the dreams of avarice.'

As trade developed, cooperages specialised. The best-quality work, wet coopering, was the making and repairing of beer and water casks. Other wet coopers made wine and spirit casks for the considerable trade in butts for rum and larger pieces with a capacity of up to 500 gallons (2273 litres). This work was shared between the *raiser*, who dressed the staves, raised and fired the casks, and made the hoops, and the *header*, who chimed the cask, made the heads, and finished off the cask. In the eighteenth century English coopers went to Portugal to teach the Portuguese how to make port pipes.

Poorer-quality work was the province of the dry cooper, sometimes called a dry bobber, who made casks for fruit, vegetables and all manner of hardware. Then there was the more demanding *dry tight* work, the making of casks for fish, syrups, oils and tar. A cooperage specialising in white work made straight-sided vessels, churns, small vats, buckets and bowls of a very high quality. Other specialist trades linked to coopering grew up. In Bristol hoopers were a separate trade at one time, just making and fitting wooden hoops, but in the eighteenth century the management of beer and wine cellars and the management of storehouses for imports were regarded as branches of coopering.

With the advent of steam power in the nineteenth century came steam-powered machines. Band-saws were among the first and were welcomed as they did the laborious, unskilled work. A considerable number of machines were required to make a cask and therefore coopering was slow to be mechanised. Since beer casks can last from thirty to fifty years in use about ninety per cent of brewery coopers' time was taken up with repair work. Machines could not repair casks. One large independent cooperage, Shuters, Chippendale & Colyers, which employed over 650 coopers, installed machinery but overextended itself and became insolvent.

By the twentieth century, as other markets closed, cooperages like Shaw's of Poplar, Wilson's of Bermondsey and Kilby's of Banbury turned to making brewers' casks. Hall's cooperage in London imported used bourbon casks from the United States and resized them for Scottish whisky distillers. Coopering by hand still seemed a secure occupation – so much so that when the author started his apprenticeship at the beginning of the Second World War he was assured, 'They've always had barrels. They'll always need coopers.' Alas! How wrong they were. The output of canned and bottled beer expanded enormously after the Second World War. As quartered oak was scarce and expensive, and cheap, lined aluminium and stainless steel drums were readily available, wooden casks began to be phased out in the larger breweries between about 1950 and 1970. The situation was compounded by the introduction of pressure beer, which made cellar work at the pubs much easier and virtually eradicated any possibility of beer becoming contaminated.

Naval coopering

In the Royal Navy the ship's cooper, 'Jimmy Bungs', worked in conjunction with the victualling steward. He unheaded provision casks and looked after the spirit, wine, beer and water casks. He also kept mess utensils, ship's buckets and other coopered vessels in repair. He would seldom use new wood but would knock down any cask he could get hold of to make smaller staves for remaking and repairing.

Such mundane matters as the reuse of empty casks were of concern at

'The Ship's Cook', 1800, a watercolour by T. Rowlandson in the National Maritime Museum. The tub would appear to be a cask cut in half, a common practice and an easy way of producing a large steep tub to soak salted meat. Kilby's cooperage made them 30 inches (76 cm) long, 31 inches (78.7 cm) across the bottom head and 38 inches (96.5 cm) across the open top.

Coopered vessels from the Viking ship excavated at Oseberg, Norway. The bucket with the elaborate handle fittings, obviously funerary, was made in Ireland.

the highest levels of state. In 1513 Cardinal Wolsey wrote to Admiral Lord Howard on board the *Mary Rose* at Plymouth:

As he must be victualled for six weeks it will be impossible to provide for his revictualling unless foists [empty casks] be not more plenteously brought from the Navy to Hampton, instead of being wastefully broken and burnt. Some ships ten weeks ago received 756 pipes and have re-delivered scarce 80 foists of these. This appears to have been done by some lewd persons that would not have the King's Navy continue any longer on the sea. Orders should be taken that the offenders be punished, otherwise it will lead to the failure of the enterprise and the Admiral will be blamed.

(Letters and Papers, Foreign and Domestic, 1509–16)

The Cardinal knew that, without casks, navies could not sail. No doubt the coopers were sometimes guilty of acquiring some (the better ones) of these casks.

These casks would have held small beer, which sailors called 'swipe'. It was safer to drink than water although, as casks were not then shaved smooth inside, it is very doubtful they would have been sterilised effectively. Unless the beer had an alcohol content sufficient to kill the bacteria that could lurk in the rough crevices inside the casks, the beer would very soon turn sour. Small beer could not have been weak. In times of war ships were brought into commission very quickly and the Navy found it more difficult to provide sufficient casks. Buying casks abroad proved very expensive. In the nineteenth century the seaman's daily allowance of 1 gallon (4.5 litres) of small beer was reduced to 2

The grog tubs Kilby's made were 19 inches (48.3 cm) long with a head in the bottom of 25 inches (63.5 cm), being larger than the open top which was only 19³/₄ inches (50.2 cm) across. Years ago a seaman's ration was a gill or noggin, which was a quarter pint, diluted with four parts of water to one of rum. Sometimes lime was added to the water. As a punishment sailors would be given grog diluted with six parts of water per one of rum. When sailors thought the grog tasted weak they would accuse the steward of 'drowning the miller'; if they had short measure it was 'plush'. Some sailors would collect their grog in a 'monkey', which was a small kid, or pot. Others might save it up in a bever barrel so that they could have a morning glass, a 'cauker', or, at midday, a 'declination'.

Washing tubs were made in three sizes with stave lengths of 10, 9 and 8 inches (25.4, 22.9 and 20.3 cm) respectively. They had 18, 16 or 15 inch diameter head (45.7, 40.6 or 38.1 cm) and the open ends were 20, 18 or 17 inches across (50.8, 45.7 or 43.2 cm). Sailors called them jolly boats.

Left: *Funnels were made 7½ inches (19.1 cm) long, 13 inches (33.2 cm) in diameter in the head and 16½ inches (41.9 cm) across the open end.*

Below: *Fire buckets were 12¼ inches (31.1 cm) long with 15 inch (38.1 cm) lugs, 7½ inches (19.7 cm) across the bottom head and 10½ inches (26.7 cm) across the open end. Ordinary buckets were in two sizes: the large ones were 12 inches (30.5 cm) long with 15 inch (38.1 cm) lugs, a 11½ inch (29.2 cm) diameter head and a 14 inch (35.6 cm) open end; the smaller bucket had 10 inch (25.4 cm) long staves with 12 inch (30.5 cm) lugs, a 10 inch (25.4 cm) diameter head and a 12½ inch (31.8 cm) open end.*

quarts (2.25 litres) of stronger beer to save cooperage. Later, as the purity of water could be relied upon, the beer allowance was replaced by grog.

Economy was still the watchword in cooperages in the nineteenth century, and on piecework coopers would receive more pay for using old staves and heading rather than new wood. Article 12 in the manual of *Cooperage Instructions* reads:

> …the Master Cooper is to take great care that new scantling and old staves are manufactured into the casks for which they will best answer, so as to be worked up to the greatest possible advantage; that the old staves which may be in his charge are to be worked up into casks…in preference to using new timber…and that he is never to convert new staves into scantling for new casks when he may have old staves which can possibly be applied to…

Coopers would have quite a stack of old staves by their berth, hoping that their length and the distance between top and bottom grooves, as well as the amount of height (bend) would coincide with the cask they were repairing. They would then only have to be jointed – and the side staves jointed – before being inserted.

Each of the first four ships of the line of battle had four coopers; a fifth line-of-battle ship had three coopers; a sixth, two; and a sloop, one. In 1861 the Admiralty gave instructions for every ship newly commissioned to have a set of cooper's tools, with a larger set for foreign service. The set consisted of a bick iron, a bolster (chince), a bung borer, chalk, a vice, a driver, extra helves (hafts or handles), flag, a gimlet, marking irons, a punch, rivets and a striking hammer. Coopers supplying their own tools received an extra twopence a day.

The ship's cooper was not expected to make any casks. One of his most important tasks was looking after the water casks. Until the nineteenth century it was the cooper's duty to *binge* (scour or rinse out) the *gang* casks that held the drinking water for everyone on the ship. The water would soon become putrid, or *addle*, in a cask that had not been scoured properly. It was not uncommon for tap holes to be bored higher in the head to avoid the sediment. He would also have to prepare a funnel to be taken ashore for filling the gang casks. The *bombard* was a very large water cask that was also used as ballast. The *leaguer* was a large water cask of 108 gallons (491 litres), equal to one butt or pipe or two hogsheads. From the mid eighteenth century the Navy required all its water casks to be iron-bound, despite the extra cost over wooden-bound casks.

Harness casks were made in various sizes to hold from 1 to 3 cwt (about 50–150 kg). They were straight-sided with the larger head at the bottom. The top head was hinged with a hasp and staple so that it could be locked. Harness casks held all manner of provisions safe from rats and thieves.

Coopered vessels used by a mess aboard a nineteenth-century man-o'-war.

On the gun decks of the old men-o'-war were casks of 'fighting water' for the gun crews. These casks were usually *scuttle butts*, which had a large square cut out at the bung through which a mug could be inserted. A little vinegar was added to the water to discourage excessive drinking. Water casks on lifeboats were made oval to fit under the thwarts (seats) and were called *barricoes*, although sailors called them *breakers* (from the Spanish word for barrel, *bareca*). They had large bungholes so that a mug could be inserted for allocating a measure of water to each of the survivors. Large power boats carried aboard battleships and heavy cruisers would have six 10 gallon (45 litre) barricoes. Smaller power boats carried four 6 gallon (27 litre) barricoes, as would cutters, whalers, 27 foot (8.2 metre) surf boats and flat-bottomed landing craft.

For over four hundred years coopers' shops could be found in victualling yards throughout the world. During the First World War Chief Petty Officer 'Teddy' Knight was the Master

A barricoe for holding water. Copper hoops have replaced the usual iron ones.

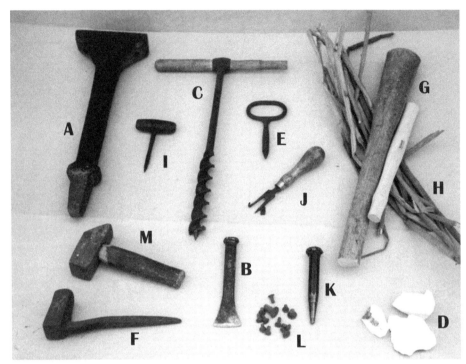

A set of ship's cooper's tools (1861 Admiralty Instructions): A, bick iron; B, bolster or chince; C, bung borer; D, chalk (for helping hoops to stick when being driven); E, heading vice; F, driver; G, helves; H, flag; I, gimlet; J, marking iron (race); K, punch; L, rivets; M, striking hammer.

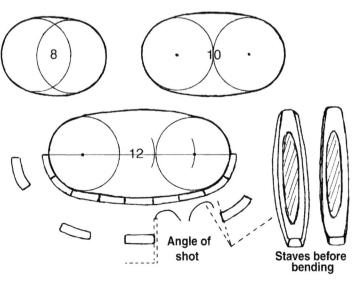

Oval casks are made in three different shapes. All the staves have to be jointed with different angles of shot.

Right: *A lade gorn or lade bucket for lading water (bailing out water).*

Cooper in Malta, a cooperage that closed on 31st March 1964. In Britain civilian coopers ran the cooperages. Clarence Yard in Gosport once employed over fifty coopers; by 1964 this number had dwindled to five, which constituted a third of the total number employed in naval yards. In charge at Gosport was the 'Leading Man of the Coopers', subordinate to the Master Cooper at Plymouth. At that time coopers were mostly employed making half hogsheads, kilderkins and firkins from larger casks, for rum distribution aboard Her Majesty's ships.

In earlier days all manner of ship's and mess utensils were made by coopers and such was the demand that tenders were sent out to independent cooperages. Kilby's cooperage at Banbury supplied thousands of tubs, buckets and casks for the Navy. Among them were naval bread barges (which were required to be 20³/4 inches [52.7 cm] long, 16 inches [40.6 cm] in diameter at the bottom and with a 13 inch [33.0 cm] open end) and large spit kits (spittoons, required to be 8 inches [20.3 cm] long, 16 inches [40.6 cm] in diameter at the bottom and 18 inches [45.7 cm] at the open top). For the more accurate spitter small spit kits were provided (6 inches [15.2 cm] long, 12 inches [30.5 cm] in diameter at the bottom and 14 inches [35.6 cm] at the open end).

At the end of the seventeenth century a cooper's pay in the Navy was £1 4s a month, on a par with an able-bodied seaman, a cook's mate and a coxswain's mate. To supplement their pay, coopers would make small tankards for members of the crew. Much earlier the cooper on the *Golden Hind* made tankards 'for such of the company who would drink from them' from the gallows on which Ferdinand Magellan had hanged mutineers. Cooper-made tankards were found on the wrecks of the Swedish ship *Vasa* and the *Mary Rose*. The Navy was obliged to increase the pay of coopers at the beginning of the nineteenth century in order to attract better men. A master cooper received £2 5s 6d a month, a cooper's mate £1 18s 6d and a cooper's crew £1 15s 6d.

Above: *Pump cans held up to 3 gallons (13.6 litres) of water.*

A tankard from the Swedish ship 'Vasa'. The spout had grown naturally and been cleverly incorporated in the tankard.

Buoys

Of the thousands of cask staves that were brought up to the surface from the wreck of the *Mary Rose*, which sank in 1545, some dozen or so caught the attention of Howard Murray, who was in charge of conservation. He asked the author if he could raise them up into a cask, which he did. They turned out to be an anchor buoy carried aboard ship for the purpose of tying to the anchor rope so that the anchor could be retrieved in the event of the ship leaving port in a hurry with no time to weigh anchor, and also for dislodging an anchor on the sea bed when weighing anchor.

Isaac Cotterel, a Bristol cooper writing in the eighteenth century, described how these buoys were made and illustrated his explanation with drawings. There were two sizes, puncheon and barrel, and they were made from tough, stringy American red oak. On completion they were tested for being 'windproof'. Sometimes they were fitted with pointed ends, which slid over the chimes and were fixed with leather, glued and painted with white lead.

An anchor buoy was part of the equipment of British and French warships. In 1800 the Navy had six standard types of buoy, the first rate costing £1 13s 1$^{1}/_{2}$d, and the sixth rate 15s 0$^{1}/_{2}$d. The Dutch bowl float was a type of cooper-made buoy that could fly a flag or pennant.

Outside the Bembridge Maritime Museum on the Isle of Wight stands a large conical marker buoy made by coopers out of American red oak. It

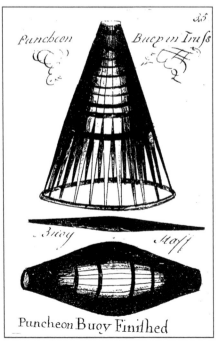

The author putting together staves of an anchor buoy that had been buried in the silt of the Solent for over four hundred years.

Left: *A drawing made by Isaac Cotterel of Bristol in 1764, showing a buoy raised up ready for firing (bending). The bend is all in the pitch, so the hoops in the pitch need to be kept very tight throughout the firing to avoid 'ducks' (broken staves). It would be necessary for two men to hammer together at exactly the same time on opposite sides of the buoy when they are firing, and the hoops would have to be dry and thoroughly chalked so that they grip, because with the angle of splay on this buoy hoops would tend to slip.*

Far left: *A conical marker buoy retrieved from Poole Harbour in 1977. It is now at the Bembridge Maritime Museum, Isle of Wight.*

Left: *A Dutch bowl float that could fly a flag in order to be recognised more easily.*

stands about 8 feet (2.4 metres) high and is about 5 feet (1.5 metres) in diameter at the base. It was retrieved from Poole Harbour in 1977.

Cable buoys were specially made to hold cables and ropes away from where they might foul and fray on rocks on the bottom. Sometimes ordinary casks were used for this purpose.

To estimate the weight in pounds that an empty cask will support in water, the capacity of the cask in gallons is multiplied by ten. For the *safe* buoyancy, the capacity is multiplied by nine, so that the safe buoyancy of a 3 gallon cask is 27 pounds; a 9 gallon firkin, 81 pounds; a 36 gallon barrel, 324 pounds; a 54 gallon hogshead, 486 pounds; and a 72 gallon puncheon, 972 pounds.

A small mooring buoy with a capacity of 3–4 gallons (13–18 litres).

Coopers of the Eltwater Gunpowder Company show off the powder kegs they are making.

Gunpowder

Most gunpowder manufacturers had their own cooperages and coopers made up about half their workforce. The making of gunpowder in Europe goes back to the fourteenth century. The demand for blasting powder grew with the economy but, for military and naval purposes, output expanded rapidly in time of war and contracted equally rapidly upon the outbreak of peace. Under these circumstances it was difficult to get the best coopers.

A barrel of gunpowder held 100 pounds (45 kg) and was called a number 1, a 50 pound (22.5 kg) powder cask was a number 2, and so on (the figure being the number required to multiply the capacity in pounds to produce the number 100). The smallest, a 30, held just 3 pounds 5 ounces (1.5 kg). Isaac Cotterel, writing in

Gunpowder kegs were made of stout oak, with very little height (bend). The wooden hoops were at the chimes, none near the small bung-hole. The heads, the weakest parts, were sunk into the cask so that it was said that more powder could be held on top of the heads than between them. They were numbered according to how many there were to a 100 pound barrel. A number 2 was a 50 pound powder keg; a number 4 held 25 pounds; an 18 held just over 5½ pounds.

Left: *A privateer's match tub. It is similar in size and shape to a horse bucket but with an open top and five notches cut into the top ends of the staves. The gun's crew would lay their match across the top of the tub between two notches, ready to touch and discharge the cannon.*

Below: *The budge barrel was another kind of powder keg. It was copper-hooped and had a leather tie-up instead of a head.*

the eighteenth century, describes the powder barrels produced for ships in Bristol cooperages as

> ...very thick, and straight in the bulge ... and extraordinary deep chimes ... they are of such a depth that between the heads will not hold more than a top of the head. There are no hoops on the bulge, but the larger ones might be hooped nearer the booge according to the cooper's discretion.

The head was the weakest part of the cask and was therefore protected by being sunken. The hoops were some way from the bung-hole so that, if a spark ignited powder trapped under the hoops, it would splutter away from the bung-hole. A cannonball striking a powder keg would have been more likely to knock it away than cause an explosion. Interestingly there is another connection between coopering and cannons. The first cannons were made by hooping together metal bars to form the tube in the same way that a cooper raises a cask, hence the business part of a cannon was called a barrel.

Maritime coopering

'The Celts are fine coopers...their casks are larger than houses,' wrote Strabo, the Roman historian. Writing about Celtic boats and the rougher seas they had to encounter, Paul Johnstone speculated that '...the simpler but strong techniques of the cooper may have been of greater value than the more elaborate mortice and tenon methods of the Mediterranean shipwright'.

As larger sea-going ships were built, the correlation between casks and ships was such that the tonnage of a ship was calculated on the number of tuns it could carry. A tun was a large cask equal to two butts or pipes and would have weighed much more than one ton.

A bas-relief from Cabrières d'Aygues, now in the lapidarium at the Musée Calvet in Avignon, France.

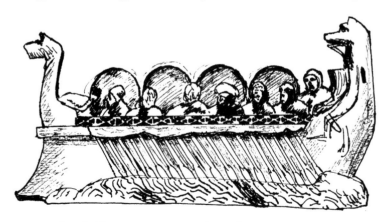

A bas-relief from Neumagen, now in the Landesmuseum in Trier, Germany.

A drawing made by Isaac Cotterel in 1764 of a butt undergoing repair. For the benefit of cooper apprentices, he explained how it was done, and it was still being done in the same way in the twentieth century. The bottom head is left in position, whereas beer casks are repaired the other way round, with the head at the top because their grooves are much deeper. New staves to be inserted are jointed and made to fit tightly in position next to the side staves, which are similarly jointed. A rounding adze is used to match up the new staves with the old. In the eighteenth century a jack plane was used. A rining adze levelled the inside of the chime before the groove could be cut with a croze. The head was then removed from the other end, which was chimed in the same way. The heads could then be fitted and the joints levelled before the hoops were driven on permanently.

Storage

Storing casks aboard ship involved several procedures. 'Strike down!' was the order for casks, held by *can hooks* (chime grips), to be lowered into the hold, where they had to be firmly stacked and secured. A rolling cask weighing half a ton could, like a loose cannon, do a lot of damage.

The tools used to repair butts in the eighteenth century: A, rining adze; B, rounding adze; C, croze; D, jointer.

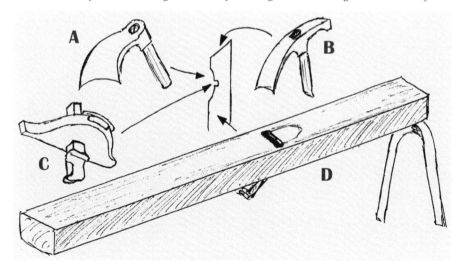

For this reason they were often held by chocks, *bilge free*, that is clear of the deck, and with quarters supported.

All casks were stacked bung uppermost. *Longer* rows were at right angles to the direction of the ship while casks *a Burton* were in line. Smaller cask storage areas were called *fagots*, where casks might well be stacked *bilge and chime* (belly to end) or *chime and chime* (end to end).

Sleepers were casks on the ground tier. *Riders* were casks piled in higher tiers. The Navy stipulated in manuals as late as 1951 that each riding cask should be supported by four casks in the tier below. The maximum number of tiers of barrels should be eight; of hogsheads, six; of puncheons, four; and of pipes and butts, three. This was to prevent the bottom casks from being stove in. In earlier times the height of the ship's deck governed this.

The *cont line* was the space between the bilges of two casks stacked side by side, and the *cut line* separated casks stacked end to end. The rearranging of casks aboard ship was called *cutting*. Casks stacked in racks were kept rigid with *dunnage* (pieces of waste timber). Single casks would be *bedded* in dunnage.

Dangerous fumes can build up in spirit casks. The sinking in Caledonia Bay of the ship *Olive Branch*, which was bringing supplies to poverty-stricken Scots, was probably caused by the cooper inspecting a cask with a naked light and accidentally setting fire to the accumulated brandy fumes.

The slave trade

During the eighteenth century Bristol coopers were kept busy making casks for the slave trade. A *negroes' crew* had a capacity of 10 quarts (11.4 litres) and was specially made to be the eating utensil for ten slaves. It was 7 inches (18 cm) deep, almost straight-sided, and measured 10 inches (25 cm) across the bottom head. It had two iron hoops, each hoop pinned to the vessel by three wooden pegs. Negroes' crews were made strong so that the slaves could not pull them apart.

Puppu tubs were made about 15 inches (38 cm) high with the bottom head 14 inches (35 cm) across, and bigger at the bottom than the top. They would hold about 8–10 gallons (36–45 litres) and had a wooden cover for the top. They were used for conveying what is referred to by Cotterel with characteristic indifference as 'the slaves' dung' to the upper deck to be thrown overboard.

Horse buckets were large 10–12 gallon (45–54 litre) vessels for holding water for the slaves. They were bigger

Negroes' crews. Each one was designed for ten slaves to eat from.

A puppu tub.

at the bottom than at the top, with a large hole in the middle of the head. A metal handle on either side took the rope when the bucket was hoisted out of the hold.

Bristol coopers also made *half powder barrels* to hold a hundred lead bars and 200 gallon (910 litre) *Guinea butts*. These casks, which seem to have been destined for the West Indies to be filled with rum for the third leg of the Triangular Trade, were shipped in *shooks* (the cask dismantled and the staves and heading bundled together). Since the heads were not dowelled together but left in pieces, they were cut to shape using a *shearing hatchet*, a tool that went out of use at the end of the eighteenth century.

Smuggling

Small-scale smuggling by individuals has never been considered a serious crime (at least, by those engaged in it) and, over the centuries, a considerable number of people did engage in it in one way or another. Small casks of brandy were not too difficult to conceal: what eighteenth-century gentleman would dream of searching a rather stout matron or a seemingly pregnant lady? A clever cooper might fit four heads into a cask so that, when the Excise officer came to take a sample through the bung-hole, he would find only cider and miss the brandy concealed at both ends of the cask.

A shearing hatchet.

There was also large-scale smuggling, when casks of brandy and other contraband were brought over from France and dropped on to the sea bed close inshore. The casks were usually tied together. At night, when (literally) the coast was clear, smugglers would go out in small boats to retrieve the shipment, having located it through a *peep tub*, a bucket-shaped tub with handles and a glass head which was used from a small boat to inspect the sea bed from beneath the surface.

There have always been men determined to get at the contents of a cask and to leave no evidence of their theft. Some sailors and dock

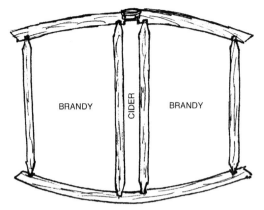

A very clever cooper might have been able to make a four-headed cask to conceal contraband.

BRANDY   CIDER   BRANDY

workers were particularly adept. They would loosen two adjacent hoops and use a gimlet to bore a small hole where each hoop fitted. Blowing through one of the holes they could catch the spirit pouring out of the other hole. After plugging the holes they would knock the hoops back over them to hide the evidence. This was called 'sucking the monkey'. The author came across this in the 1950s on a brandy butt in a bonded railway wagon. Nelson's body was brought home from Trafalgar in a leaguer, a large water cask, filled with brandy. It is said that two seamen aboard the ship bringing his body home were found the worse for drink and, on being questioned, confessed that they had been 'tapping the admiral'.

Submarine and diving vessels

In the early eighteenth century a number of men were lowered in cooper-made casks on to wrecks in search of

Left: *A peep tub, with glass head, used by smugglers to locate casks of contraband left on the sea bed.*

Small casks like these might have been used for smuggling.

Above: *The sort of tools used by eighteenth-century coopers: A, dog (used by dry coopers for levering hoops); B, bow screw compass (usually with wooden threads which had to be soaked periodically); C, rining adze; D, measuring stick (for sawing off staves to length); E, chime saw (for sawing grooves in individual staves); F, riggle gauge (for measuring the distance from the end of the chime to the groove); G, sheering hatchet (used for cutting in large undowelled heads for sending away in shooks); H, scotching hatchet (used for firing small casks); I, rushing iron (preceded the flagging iron).*

Coopers at work in the eighteenth century.

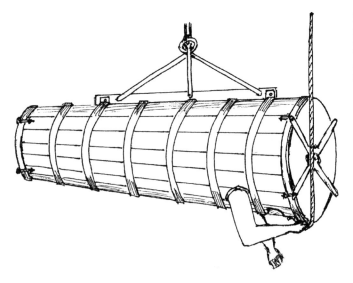

The author's impression of Lethbridge's 'Engine', based on an article in 'History Today' and a drawing of 1730 in the Shetland Museum.

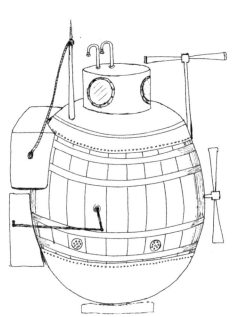

treasure. John Lethbridge, who worked between 1715 and 1759 at depths of up to 10 fathoms (18 metres), was highly successful. A cooper in Stanhope Street, London, made him a straight-sided cask, 6 feet (1.8 metres) long, 2¹/₂ feet (76 cm) in diameter in the head and 18 inches (45 cm) at the foot, and iron-hooped inside and out. Lethbridge retrieved almost £100,000 in coin, silver bars, cannon and slabs of lead from a number of wrecks across the seas (Zelide Cowan, *Port of London*, 1979, second edition).

During the American War of Independence a colonial cooper named David Bushnell built a submarine from a large wooden cask in the coopers' shop of Joseph Borden at Bordentown, New York. He named it *The Turtle*, because it reminded him of two turtle shells put together. Propelling paddles were turned by hand from inside the cask. 200 pounds (90 kg) of lead ballast could be jettisoned from inside and brass pumps could force water ballast out of the cask. Also fitted to the

The author's impression of 'The Turtle', based on the observations of Bushnell's contemporaries and the inventor's description.

'King Alexander's Submarine', from the 'Romance of Alexander', a manuscript written and illustrated in Flanders in 1340 and now at the Bodleian Library, Oxford. Maybe this was the first submarine.

outside of the cask was a time-charge of gunpowder, which could be screwed into the bottom of a ship. *The Turtle* never did sink a ship, perhaps because the copper sheathing on the hulls of the British ships was too hard, but the cooper's cask takes its place as the first submarine warship.

Work in the docks

Casks were inspected by a cooper before leaving the ship and marked for the attention of coopers ashore, who were kept busy repairing leaks as soon as the casks were on the dock. Cooperages sprang up around the docks and had a steady supply of work with casks. In Bristol in the eighteenth century master coopers reckoned that storehouse work at the docks was more profitable than

In the mid eighteenth century Bristol cooperages supplied ships with huge numbers of casks. This is an invoice from Cotterel's cooperage for casks supplied to the privateer 'Success', dated 24th September 1768.

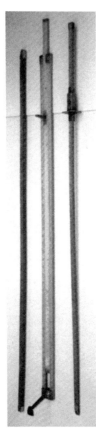

In the centre is a head rod, one of a number of gauging instruments used for calculating capacities. To each side is a dip stick used for measuring through the bung-hole or tap hole upright and diagonally.

making casks: there was the repairing of casks, checking capacities, unheading and sampling, and working in conjunction with health inspectors and Customs and Excise. Coopers needed a number of different dip sticks and measuring rods for this work.

Casks with the same capacity but made at different cooperages might vary considerably in shape. Some might have far more height (curvature) than others and some would vary in length. It was therefore quite difficult to measure exactly the contents of a cask holding less than its full amount. The capacity of a cask in cubic inches could be found mathematically by adding two-thirds of the difference between the diameter at the head and the diameter at the pitch (belly) to the head diameter, squaring it, and multiplying it by 0.7854 and then by the length. To find the capacity in *ale gallons* they would multiply twice the diameter at the bung squared plus the head diameter squared by the length and then divide by 1077. Dividing instead by 882 would give the capacity in *wine gallons*. The results would probably have been more accurate if the cask could have been weighed empty and then full of water, since water weighs 10 pounds per gallon.

Before 1824 the British wine gallon was 231 cubic inches. It was called the Winchester wine gallon and is still the standard gallon in the United States. The British ale gallon was 281 cubic inches. The Imperial gallon is slightly smaller, at just over 277 cubic inches.

In the eighteenth century coopers in the London docks were specialising: dry coopers on tobacco and sugar, and wet coopers – the more skilful – on rum casks, of which vast quantities were being imported. Many coopers worked in the bonded warehouses and vaults, where fungus grew to surprising lengths over everything. The cooper could be heard sounding out casks with his *flogger* (a type of long-handled mallet) to ascertain whether any cask had leaked.

In the twentieth century the number of coopers employed in the docks declined. Container ships and the use of large metal and plastic containers to transport wine and spirits meant less work for them to do. They could still be called on to travel from St Katherine's Dock to the Surrey Docks and to the Victoria and Albert Docks, or perhaps to Tilbury, to perform general work on all kinds of casks, wet and dry, but by the 1970s all coopering in the docks was finished.

Fish

The fishing industry and coopering were closely associated. Coopers made vast numbers of casks for fish in medieval times. Salmon was exported from Aberdeen during the thirteenth century and eels were also transported in casks. Coopers made shallow casks for oysters called *oyster bars*.

Large numbers of casks were made for the whaling industry, some for whale oil and others for blubber, which was cut into strips in order to force it through the bung-hole in a process called 'making off'. Wilson's cooperage in Bermondsey was one of a number of cooperages around Britain kept busy making casks for whalers.

The herring industry was very important. Huge herring shoals came

A stack of herring barrels, surplus to requirements.

down the North Sea during the spring and summer and coopers followed them from Peterhead down to Great Yarmouth, heading up casks of salted herring. For the other six months of the year they made the casks, which were of slight, soft wood. The insides of the chimes were sloped so that the head could be removed and reinserted quite easily. A *flincher* was the tool used by herring coopers for this job. The work was called *dry/tight* coopering. By 1913 fifteen factories were turning out over a million casks a year and 1500 coopers were making them by hand. The trade declined tremendously between the wars and by the 1960s was almost finished, as were the herring from over-fishing. Most of these salted herring were exported to Russia in the so-called Klondyking trade.

Miscellany

In the cool chalk caves of Champagne Mercier at Epernay in France stands an enormous cask with a capacity of 160,000 litres (35,000 gallons). It was made by Monsieur Jolibois, the Epernay cooper, and took thirty-one years to construct. For the oak he went to Hungary, where, with the help of a Hungarian cooper named Walter, he selected two hundred and fifty oaks, had them felled and quartered, and then dressed and bent the individual staves by the side of Lake Balaton, where, tragically, Walter was drowned. Jolibois worked on alone. With scaffolding the height of a three-storey house, he fitted the staves together, levelled them with an adze and chimed the vessel. With a pair of compasses 8 feet (2.4 metres) in radius he measured and cut in the heads. Six years after it was completed, in 1889, hauled by twenty-four white oxen and cheered all the way by huge crowds, it arrived in Paris for the Great Exhibition of that year.

In the cellars of the castle at Heidelberg in Germany can be found an even larger cask, with a capacity of 221,726 litres (48,780 gallons). Older than the one at Epernay by more than a century, it was raised in 1751 by Kufermeister Englert. His huge compasses and 5 foot (1.5 metre) *short-jointer* hang from the wall. Astride the great vessel is a balustraded dance floor but it is the eighteen stout wooden hoops, each over 9 inches (23 cm) thick, that make the tun so impressive. It was filled twice by the peasants' taxes, paid in wine.

In England the largest cooper-made vessel on record was a colossal vat, raised at the Meux brewery in Clerkenwell, London, in 1806. It was 44 feet (13 metres) in diameter and 30 feet (9 metres) deep, with a capacity of 272,520 gallons (1,238,868 litres), and was made to hold porter, which needed to mature for twelve months. A considerable number were built until one burst its hoops in 1814, drowning or suffocating eight people. After that disaster the maturation time for porter was shortened dramatically so that smaller vessels could be used.

The Great Mercier Cask on its way to Paris.

The Heidelberg tun, raised in 1751, supports a dance floor.

Beneath the great tun at Heidelberg, the wooden hoops, each 9 inches (23 cm) thick, hang loose.

Two types of bever barrels: some sailors used them for their grog.

During the eighteenth century coopers in Bristol were making *bottles* for Scottish distillers to hold samples of their liqueurs sent to England. Afterwards they were used as *bever barrels* on the farms or by seamen for their grog. Yet in Scotland at this time more elaborate ornamental bottles called *bickers* were being made. With alternating light and dark woods, the joints were *feathered* as on the smaller brass-hooped bottle and the *quaich* (whisky mug) in the photograph below.

Until corks became plentiful at the end of the eighteenth century and there was more understanding of how wine could mature and acquire a bouquet in the bottle, wine was sold in casks and cellar work was a

Three bickers (ornamental bottles) and a quaich (whisky mug). The bicker on the right has been feathered.

A 3 gallon (13.6 litre) softwood cask, referred to as a bucket, with two holes in the top head, was used for taking hot and cold sea water to customers at Ventnor on the Isle of Wight in Victorian times. The water was thought to have therapeutic properties. Although they represented a considerable weight, a man would carry two of these hung on a yoke. (Information from Ventnor Heritage Museum.)

profitable branch of coopering. 'Our wine coopers of recent times use vast quantities of sugar and molasses in all sorts of wines to make them drink brisk and sparkling,' wrote Christopher Merit in 1662 in a paper read to the Royal Society. Sparkling champagne is mentioned in a play written by Sir George Etheridge in 1676. This preceded Dom Perignon's claim. By the nineteenth century cellar work of this kind, as a branch of coopering, had come to an end. What cannot be taken away from the trade, however, is the cooper's claim to be 'the man who invented champagne' (Simon Trump, *Mail on Sunday*, 12th July 1998, quoting Tom Stephenson).

Nevertheless, it is that vast army of coopers engaged in their humdrum and mundane daily tasks who won the greatest accolade. It can be found in Henry Mayhew's *London Labour and the London Poor*, where he quotes a street hawker selling penny broadsheets depicting the gory details of a murder: 'When a woman is bad, she is bad...there's the board...the 'llustration, it seems to cooper the thing. They must believe their eyes.' And, bemoaning the fact that some murders failed to fire the public imagination, Mayhew wrote: 'Why, there was William Gleeson, as great a villain as ever lived...went and murdered a whole family. But Rush coopered him. Made it no draw to anyone.' There can be no finer praise than when the name of a trade enters the vernacular in so complimentary a manner. It seems to cooper it.

One other type of cask is worth mentioning here. At funerals in Vietnam a barrel drum made in the Ancient Egyptian tradition is played. The very slight staves are bent over a bar heated on a small stove. After being planed into shape they are raised up in iron hoops. Buffalo hide is stretched over the ends and held firm with bamboo pegs. For ceremonies they are painted bright red. The one in the photograph was left unpainted by request (Laurie Sayer, *Tool and Trades History Society Newsletter*, 63).

The Vietnamese drum, played at funerals.

Coopering today

In England most of the independent cooperages – Shaw's, Wilson's, Kilby's, Oldham's and many others – closed down in the 1950s. Perhaps the last to go out of business was Buckley's of Manchester in the early 1990s after producing a number of 150 gallon *Burton Unions* for Marston's brewery to brew a high-quality Pedigree beer. A few breweries – Samuel Smith's, Theakston's, Wadworth's and Marston's – still employ a cooper, but this is mainly for public relations and advertising purposes; and they still maintain a number (albeit ever-decreasing) of old casks. Hall's of East London, which went out of business as late as 1984, used to remake American bourbon casks into whisky hogsheads for Scottish distillers. They also exported to Japan. Four of their coopers continued making ornamental and fancy casks for a small discerning market.

There are still a few coopers working in Scotland, preparing empty sherry casks for the better-quality whiskies.

When, in the 1970s, some wines began to be imported in large, stainless steel and plastic containers – overcoming the storage and handling difficulties inherent in wooden casks – it seemed that the days of the barrel were coming to an end. However, this is not the case: in some countries mechanised coopering has continued to thrive. Wines can derive considerable benefit from maturing in oak casks and, whereas

Cooper and Braumeister Craig Dunn of the United States stands beside a stave-jointer machine in his mechanised cooperage.

Gone is the skill and craft of the old cooper; now the machine does the work at Craig Dunn's cooperage in the United States.

this used possibly to have been regarded merely as a bonus, now it is the reason for the continued production of casks. Today there are, broadly speaking, two branches of the trade. First, there is the spirit branch: in the United States, about 1.3 million bourbon casks are produced yearly and, in cooperages in south-east Asia and Japan, the output is about 200,000 casks. These casks are mass-produced and are of a relatively poor quality. Second, there is the making of wine casks, *barriques*, of 225 and 300 litres, in the French tradition. French cooperages are now producing over half a million barriques per annum, a considerable number of them

'Toasting' casks at Remy-Martin in France. These casks are slightly tilted so that the draught keeps the fires burning merrily, browning the insides of the casks and releasing the flavours of the oak. Toasting is specified as light, medium or heavy.

for export. These casks are extremely well made and great attention is paid to the *toasting* that is carried out after the cask is fired. Much scientific study has gone into the maturing of wine and spirits in oak. For toasting, the cask is put over a cresset filled with oak shavings; radiant heat browns the interior wall of the cask, developing a caramel flavour from the oak hemicelluloses, as well as a vanillin flavour from the oak lignin. In Australia the coopering industry has been growing almost continuously over the past two decades, while in the Napa Valley of California a great expansion of activity in coopering took place in the 1990s.

The approximate number of casks used per annum in the world's wine industry is shown in the following table.

Wine-producing area	Number of casks
France	300,000–400,000
California	250,000–300,000
Australia	110,000
South America	50,000
Spain	50,000
Italy	30,000

Oaks are not grown in the southern hemisphere, so that Australian cooperages have to import all their timber (mostly from the United States and France, and some from northern Europe). These cooperages have an excellent reputation for producing some of the finest barriques, superior to the French and American casks. Barriques have a working life of about eight years. Usually for the first two years they contain a superior wine; during the next two years they hold wines of lesser quality, after which they may be used for fortified wines and then perhaps as flower tubs.

The Canadian Niagara wine industry used to import American white oak barrels but it has started to use Canadian white oak, which gives more of a vanilla flavour to the wine than the white oak grown in more southerly latitudes and has similarities to French oak. However, there being no Canadian cooperage industry, the oak is sent down to the United States to be made up into casks by cooperages in Missouri. Yet, while the trade seems now to be dominated by large machine cooperages in wine-producing areas, it is interesting to note the emergence of a small mechanised cooperage in the eastern United States. The cooper, Craig Dunn, is a German-trained Braumeister intent on establishing a pub brewery dispensing lager beer from oak casks.

No doubt a certain amount of hand coopering will be necessary to keep casks in repair but machines will make the casks of the future. Today there are no cooperages making casks by hand commercially. Australian coopers serve a four-year apprenticeship, twenty days of which are spent at a school of further education learning the art of wood machining. The days of 'Truss oh!' and coopers teaming up to fire casks, hammering away at stubborn truss hoops amidst fire and smoke, would seem to be over. Let us hope that future generations are able to appreciate the tremendous skill of the old coopers and the backbreaking work involved in their trade.

Bibliography

Cotterel, Isaac. *The Numeration Book*. 1764. (Manuscript in possession of the Coopers' Company.)

Ditchfield, P. H. *The Story of the City Companies*. G. T. Foulis & Company, undated.

Elkington, G. *The Coopers' Company and Craft*. 1933.

Foster, Sir William. *A Short History of the Worshipful Company of Coopers of London*. 1961.

Gilding, Bob. *The Journeymen Coopers of London*. History Workshop, Ruskin College, Oxford, 1971.

Kilby, K. *The Cooper and His Trade*. John Baker, 1971, and Linden Publishing Co Inc, Fresno, California, 1990.

Salaman, R. *A Dictionary of Tools Used in Woodworking and Allied Trades*. Allen & Unwin, 1975.

The author holding a cask that he has made 'about face'. It could be used as an umbrella stand.

Glossary

Adze: *dry cooper's* – has a notch for removing nails; *rining* – for cutting the inside of the chime smooth for the groove; *sharp* – for cutting the slope on the end of the chime; *trussing* – used in driving down truss hoops when firing a cask.

Auger: *bung* – a T-shaped drill for cutting the bung-hole; *tapered* – for putting a taper on the hole; *thief* – for removing pieces of wood loose inside the bored hole.

Axe, cooper's: like the medieval side axe; for trimming straight staves and heading.

Backing: shaping a straight stave when dressing the staves.

Barrel: a 36 gallon (164 litre) beer cask.

Bever barrel: a 3-4 pint (1.7–2.3 litre) cask with mouthpiece, to hold beer for a farmer's lunch break.

Bever time: a break during work.

Bick iron: a cooper's anvil, for working with hoop iron.

Bickers: small ornamental casks made in eighteenth-century Scotland.

Bombard: a large water cask used as ballast.

Booge: the part of the cask about a third of the way down the stave.

Bottle: coopers' name for a bever barrel.

Budge barrel: a gunpowder cask with tie-up top.

Burton Union: a large cask used in series for brewing beer.

Butt: *Guinea* – a 108 gallon (491 litre) cask sent abroad in shooks; *scuttle* – a water cask with large square bung, for refreshing gun crews; *sherry* – a 108 gallon (491 litre) cask.

Buzz: a two-handled lever for a scraper blade.

Cant: D-shaped parts of a head, i.e. the sides.

Case: a cask chimed on one end for knocking down to repair other casks.

Chime: ends of staves

Chiming: shaping the ends of the staves.

Chiv: a curved plane hanging from a board. It shaves inside the chime.

Cleft: split rather than sawn when timber is converted.

Cresset: a small brazier for holding the staves when firing.

Croze: a tool for cutting the groove.

Diagonals: hinged rods used for measuring capacities.

Dingee: a metal hoop, the last to be fitted when firing.

Dog: used by dry coopers to lever staves away from the head.

Dowelling stock: a wooden brace with a fixed scoop drill for dowelling.

Downright: a two-handled plane-type tool for levelling stave joints.

Dressing staves: shaping straight staves prior to raising up.

Driver: used with a hammer to tighten hoops; formerly called a *drift*.

Dry cooper or bobber: a maker of casks for dry goods.

Duck: a stave that breaks during firing.

Feathering: interlocking staves on small casks.

Firing: bending the staves over a fire.

Firkin: a 9 gallon (41 litre) beer cask.

First runner: the first truss hoop used in firing.

Flag: the coopers' word for river rush.

Flagging iron: used for levering out staves to insert flag.

Flincher: a herring coopers' tool for cutting a bevel above the groove.

Flogger: a light mallet on a long springy haft, for sounding out casks.

Gang cask: a water cask.

Gathering hoop: an old term for *first runner*.

Gimlet: a small tool for making a small hole; held in one hand.

Gun: a cask after firing.

Harness cask: a container for 1-3 cwt (50–100 kg) of salted beef.

Hatchet: *sheering* – a large round blade for cutting undowelled heads; *scutching* – for driving down truss hoops when firing, now obsolete.

Head: pieces of wood, usually dowelled together, forming a circle or an oval, fitted into the ends of the cask. The *top head* has the tap hole.

Header: a cooper who shares the making of a large cask with the raiser.

Heading board: used for resting the head of the cask in when working at it.

Height: the amount of belly in a cask.

Hogshead: a 54 gallon (245 litre) beer cask.

Hollowing out: shaping a straight stave in the dressing process.

Horse: a wooden clamp operated by foot while sitting with both hands free.

Jigger: a one-handled knife for shaving inside the chime.

Jimmy Bungs: a cooper's nickname aboard ship.

Jointer: a 6 foot (1.8 metre) long plane used for the sides of staves and for heading.

Kilderkin: an 18 gallon (82 litre) beer cask.

Knife: *backing* – a two-handled drawknife for shaping straight staves; *heading* – a drawknife used for cutting in the head; *hollowing* – a drawknife used for shaping a straight stave.

Knocker-up: a bent bar for inserting through the bung-hole to knock the head into position.

Lade gorn: a bucket for bailing out.

Leaguer: a 108 gallon (491 litre) water cask.

Listing: shaping a straight stave with an axe.

Lord: a cask that leans, i.e. drunk as a lord.

Match tub: for holding matches for firing cannons.

Maul: a steel bar for driving down chime hoops.

Measuring stick: for measuring stave lengths for sawing before machine band-saws.

Medullary rays: impervious rays running outwards from the centre of a tree.

Memel oak: an extensively used timber exported from Russia through Memel.

Nose: part of a hammer used for splaying hoop iron; also how a cask smells (checked before racking or filling).

Oyster bars: shallow casks for oysters.

Peep tub: a glass-headed cask with two handles for inspecting the sea bed.

Piece: a very large cask of 500 or 1000 gallons (2273 or 4546 litres).

Pin: a 4$\frac{1}{2}$ gallon (20.5 litre) beer cask.

Pipe: a 108 gallon (491 litre) cask used mostly for port.

Pitch: the belly of the cask.

Pompeying: charring the inside of the cask.

Pump can: a type of jug holding up to 3 gallons (14 litres).

Puncheon: a 72 gallon (327 litre) beer cask, or one of 90 gallons (409 litres) or more for rum and whisky.

Quarter: the part of a cask between the chime and the booge.

Quartered: timber converted along medullary rays.

Riders: casks stacked upon other casks in higher tiers.

Riggle: a word used in the eighteenth century for 'groove'. A *riggle gauge* is two hinged pieces of wood for measuring the chime above the groove.

Rivet mould: for burring over the rivet.

Runlet: an 18$\frac{1}{2}$ gallon (84 litre) wine cask.

Rushing iron: for levering staves in order to insert flag; superseded by the flagging iron.

Saw: *band* – a machine saw; *bow* – a frame saw for cutting round heads; *chime* – an eighteenth-century rounded saw for cutting grooves in single staves.

Scraper: a turned blade which removes fine shavings.

Shave: *inside* – a curved plane with side handles for shaving inside casks; *pail* – a large curved spokeshave for shaving inside buckets and churns.

Shooks: casks knocked down and bundled up for export.

Shot: the angle of slope on the side of staves, corresponding to the radii from the cask centre.

Sleepers: casks on the bottom tier in a stack of full casks.

Splay: the shape of a metal hoop following the curve of the cask.

Steam bell: a steam trap for softening timber before firing.

Stoup plane: a rounded plane for shaving the inside of the cask.

Toasting: the browning of the inside of the cask over a fire.

Topping plane: a curved plane for levelling the top of the chime; also called a sun plane.

Truss hoop: an ash hoop 1$\frac{1}{2}$ inches (3.8 cm) thick for driving down to bend the staves.

Trussing: driving down hoops when firing.

Tun: a 225 gallon (1023 litre) cask.

Vat: an upturned bucket shape, made in all sizes, for liquids and powders.

Vice heading: a sharply tapered thread with hand grip for levering up to the head.

Wet cooper: a cooper who works on beer, wine and spirit casks.

White cooper: a cooper who works on straight-sided vessels.

Wind-tight: a water cask was said to have been water-tight and wind-tight.

Places to visit

GREAT BRITAIN AND IRELAND

Cider Museum, 21 Ryelands Street, Hereford HR4 0LW.
Telephone: 01432 354207. Website: www.cidermuseum.co.uk

Cookworthy Museum, The Old Grammar School, 108 Fore Street, Kingsbridge, Devon TQ7 1AW. Telephone: 01548 853235. Website: www.kingsbridgemuseum.org.uk

Curtis Museum, High Street, Alton, Hampshire GU34 1BA.
Telephone: 08456 035635. Website: www.hants.gov.uk/curtis-museum

Guernsey Folk and Costume Museum, Saumarez Park, Castel, Guernsey GY5 7UJ.
Telephone: 01481 255384. Website: www.nationaltrust-gsy.org.gg

Guinness Storehouse, St James's Gate, Dublin 8, Republic of Ireland.
Telephone: 00353 (0) 1453 8364. Website: www.guinness-storehouse.com

Hop Farm Country Park, Paddock Wood, Kent TN12 6PY.
Telephone: 01622 872068. Website: www.thehopfarm.co.uk

Melton Carnegie Museum, Thorpe End, Melton Mowbray, Leicestershire LE13 1RB.
Telephone: 0116 305 3860. Website: www.leics.gov.uk

Museum of English Rural Life, The University, Redlands Road, Reading, Berkshire RG1 5EX. Telephone: 0118 378 8660. Website: www.reading.ac.uk/merl

Museum of St Albans, Hatfield Road, St Albans, Hertfordshire AL1 3RR.
Telephone: 01727 819340. Website: www.stalbansmuseums.org.uk

Museum of Smuggling History, Botanic Gardens, Ventnor, Isle of Wight PO38 1PE.
Telephone: 01983 853677.

National Brewery Centre, Horninglow Street, Burton upon Trent, DE14 1NG.
Telephone: 01283 532880. Website: www.nationalbrewerycentre.co.uk

Ryedale Folk Museum, Hutton-le-Hole, North Yorkshire YO6 6UA.
Telephone: 01751 417367. Website: www.ryedalefolkmuseum.co.uk

Scolton Manor Museum, Spittal, Haverfordwest, Pembrokeshire SA62 5QL.
Telephone: 01437 731328.

Shakespeare Countryside Museum, Mary Arden's House, Wilmcote, Stratford-upon-Avon, Warwickshire CV37 9UN. Telephone: 01789 293455.Website: www.shakespeare.org.uk

Shipwreck and Maritime Experience, Arreton Barns Craft Village, Arreton, Isle of Wight.
Telephone: 01983 533079. Website: www.iowight.com/shipwrecks

Speyside Cooperage Visitor Centre, Dufftown Road, Craigellachie, Aberlour, Banffshire AB38 9RS. Telephone: 01340 871108. Website: www.speysidecooperage.co.uk

St Fagan's National History Museum, St Fagans, Cardiff CF5 6XB.
Telephone: 029 2057 3500. Website: www.museumwals.ac.uk/en/stfagans

Stockwood Craft Museum, Stockwood Park, Farley Hill, Luton, Bedfordshire LU1 4BH.
Telephone: 01582 738714. Website: www.luton.gov.uk

Towneley Hall Art Gallery and Museums, Burnley, Lancashire BB11 3RQ.
Telephone: 01282 424213. Website: www.burnley.gov.uk/towneley

Usk Rural Life Museum, The Malt Barn, New Market Street, Usk, Monmouth NP15 1AU.
Telephone: 01291 673777. Website: www.uskmuseum.org.uk

Ventnor Heritage Museum, 11 Spring Hill, Ventnor, Isle of Wight PO38 1PE.
Telephone: 01983 855407. Website: www.ventnorheritage.org.uk

Watford Museum, 194 High Street, Watford, Hertfordshire WD1 2HG.
Telephone: 01923 232297. Website: www.watfordmuseum.org.uk
Weald & Downland Open Air Museum, Singleton, Chichester, West Sussex PO18 0EU.
Telephone: 01243 811363. Website: www.wealddown.co.uk
Wick Heritage Centre, 19 Bank Row, Wick, Caithness KW1 5EY.
Telephone: 01955 605393. Website: www.wickheritage.org
York Castle Museum, Eye of York, York YO1 9RY. Telephone: 01904 687687.
Website: www.yorkcastlemuseum.org

OTHER COUNTRIES
Colonial Williamsburg, PO Box 1776, Williamsburg, Virginia VA 23187-1776, USA.
Telephone: 001 (0) 757 229 1000. Website: www.history.org
Hennessy Cooperage Museum, Rue de la Bichonne, Cognac, Charente 16100, France.
Telephone: 0033 (0) 5453 57268.
Le Musée Animé du Vin et de la Tonnellerie, 12 Rue Voltaire, Chinon, Loire 37500, France.
Telephone: 0033 (0) 2479 32563.
Maison des Brasseurs, Grand Place 10, 1000 Bruxelles, Belgium.
Museum voor Volkskunde, Rolweg 40, 8000 Brugge, Belgium.
Old Fort William, Thunder Bay, Ontario POT 229, Canada. Website: www.fwhp.ca

The Burton Union system preserved at Coors Visitor Centre, Burton upon Trent.

Index